Sustainable Waste Management Workshop: Microplastics in the Environment 2020

Singapore
7 - 9 January 2020

ISBN: 978-1-7138-0814-5

Printed from e-media with permission by:

Curran Associates, Inc.
57 Morehouse Lane
Red Hook, NY 12571

Some format issues inherent in the e-media version may also appear in this print version.

Printed with permission by Curran Associates, Inc. (2020)

For permission requests, please contact AIChE
at the address below.

AIChE
120 Wall Street, FL 23
New York, NY 10005-4020

Phone: (800) 242-4363
Fax: (203) 775-5177

www.aiche.org

Additional copies of this publication are available from:

Curran Associates, Inc.
57 Morehouse Lane
Red Hook, NY 12571 USA
Phone: 845-758-0400
Fax: 845-758-2633
Email: curran@proceedings.com
Web: www.proceedings.com

TABLE OF CONTENTS

POSTER SESSION

Ecotoxicity of Microplastics in Aquatic Ecosystem

Conference: Waste Management

Year: 2020

Proceeding: Sustainable Waste Management Workshop

Group: General Submissions

Session: Plenary Session I

Time: Wednesday, January 8, 2020 - 10:20am-10:45am

Authors:

An, Y. J., Konkuk University

Water environments are increasingly becoming contaminated by plastic wastes, and microplastic pollution in aquatic ecosystems is currently gaining considerable attention worldwide. In addition to the direct effects of plastic pollution on aquatic organisms, these pollutants may have potentially adverse impacts on human health, owing to the food chain transfer and substantial reliance on nutrients derived from seafood. Here, we evaluated the adverse effects of microplastics on physical, biochemical, and nutritional aspects of the major species in aquatic ecosystems. These results provide insights into the problem of microplastic pollution of the water environment and focus attention on the food chain transfer, which has direct implications on human health.

The Mechanism of PE Microplastics Carrying Cr(VI) Ions in the Water Environment

Conference: Waste Management

Year: 2020

Proceeding: Sustainable Waste Management Workshop

Group: General Submissions

Session: Session 1

Time: Wednesday, January 8, 2020 - 10:55am-11:20am

Authors:

Zhang, W., Zhengzhou University

More attention was paid to the attachment between Microplastics and environmental pollutants. The pH would influence the adsorption capability. The attachment performance of Cr(VI) by Polyethylene (PE) would be promoted by sodium dodecyl benzene sulfonate (SDBS) addition for the pH of less than 6. The attachment performance would be decreased when the pH was more than 6, for the SDBS would compete with CrO_4^{2-} for occupying the adsorption sites of PE. This research would provide a basis for investigating the influence of pH on adsorption performance of heavy metal by PE to simulate the surface attachment model of those three kind of pollutants.

Thermal-Chemical Recycling of Waste Plastics for Valuable Products

Conference: Waste Management

Year: 2020

Proceeding: Sustainable Waste Management Workshop

Group: General Submissions

Session: Session 1

Time: Wednesday, January 8, 2020 - 11:20am-11:45am

Authors:

Yao, D., National University Singapore
Wang, C. H., National University of Singapore

Plastic is rich in hydrogen and carbon resource. In year 2018, nearly 908,600 tons of waste plastics are disposed in Singapore according to a NEA report. However, the low recycling rate of plastic waste, which was only 4% that ranked the least among all kinds of waste and represented a huge waste of hydrogen and carbon resources, has been a great challenge for the sustainable development of whole society. The CO_2 emission from the current incineration process of waste plastics contributes large part in the greenhouse gas (GHG) emissions. Thermal-chemical recycling of waste plastics is a promising way to reduce waste with simultaneously conserving natural resources. In this work, waste plastic waste will be thermally vaporized (pyrolysis) firstly followed by a catalytic reforming/upgrading process, where hydrocarbon-containing compounds being converted to desired valuable products. As catalyst was very important for this process, experiments were performed into the influence of the catalyst where a series of different Ni or Fe modified catalysts were compared in relation to hydrogen yield and carbon nanomaterials production. Results found the catalyst synthesis method and the type of metal loading significantly influenced the products yields. Thick layers of carbon nanotubes, with diameters in the range of 20-40 nm and length up to tens of microns were obtained. A deep investigation on microstructure and crystalline phase of spent catalyst and products was also conducted for better understanding of the growth of carbon nanomaterials from pyrolysis-catalysis of plastic.

Microplastics Abundance, Distribution and Impacts in Tropical Urban Fresh Waters: A Singapore Case Study

Conference: Waste Management

Year: 2020

Proceeding: Sustainable Waste Management Workshop

Group: General Submissions

Session: Session 2

Time: Wednesday, January 8, 2020 - 1:00pm-1:25pm

Authors:

Choudhary, N., National University of Singapore
Zeng, Y., National University of Singapore
Mowe, M. A. D., National University of Singapore
Yeo, D. C. J., National University of Singapore

Plastic aquatic pollution has garnered much scientific attention in recent years, revealing potential impacts on biodiversity and human health. However, information surrounding microplastics in fresh waters remains scant. To address this knowledge gap, we first conducted a literature review, focusing on the global distribution and abundance of microplastics in fresh waters. Early findings from the review identify the recurring source of microplastics as wastewater effluents, indicating a spatial association between the presence of plastic with its proximity to urban areas. Finding a paucity of research in tropical freshwater systems, we followed this up with a study on selected reservoirs and canals of Singapore. Through in-situ sampling, chemical digestion, filtration, and microscopy classification, we identified the most abundant type of microplastic pollution (particle size < 5 mm) across five freshwater systems (both lotic and lentic), and suggest key factors that could contribute to this abundance. Our results represent an important first step towards understanding the impacts of microplastic pollution in Singapore, and possibly other tropical freshwater systems.

Compare Visual and Spectroscopic Identification of Microplastics in Municipal Solid Waste Samples

Conference: Waste Management

Year: 2020

Proceeding: Sustainable Waste Management Workshop

Group: General Submissions

Session: Session 2

Time: Wednesday, January 8, 2020 - 1:25pm-1:50pm

Authors:

Lu, X., Nanyang Technological University
Fei, X., Nanyang Technological University

Dumpsites and landfills contain huge amounts of spatially concentrated waste plastics for an infinite period. Waste plastics are susceptible to weathering and fragmentation which result in microplastics (MPs) that remain in the environment and almost impossible to clean up. Prior to land disposal, most waste plastics have been exposed to sunlight ultraviolet (UV) radiation, moisture intrusion, elevated temperature, mechanical stress, and may be stained by organic solvent and corrosive liquid. These exposures initiate and accelerate the weathering and fragmentation processes of waste plastics. However, the detection methods and available measurements on MPs occurrences in landfills are lacking. In this study, we separate and characterize MPs in environmental solid samples by density-based extraction and filtration followed by selective fluorescent staining using Nile Red. The dried solid samples were mixed with saturated sodium chloride solution, and the supernatant was subsequently filtered by cellulose acetate membrane filters with a pore size of 0.45Î¼m and digested with 30% hydrogen peroxide. The recovered particles were combined, filtered again, photographed and visually counted. The particles were confirmed by Fourier-Transform Infrared Spectroscopy (FT-IR). The best concentration of Nile Red is 10 Î¼g/L, the microscopic fluorescent images can be further analyzed by software. There is no significant difference between the visual and spectroscopic results of the MPs. The composition of MPs in different waste fractions of solid waste samples were detected.

Characteristics of Particulate Plastics in Terrestrial Ecosystems

Conference: Waste Management

Year: 2020

Proceeding: Sustainable Waste Management Workshop

Group: General Submissions

Session: Session 2

Time: Wednesday, January 8, 2020 - 1:50pm-2:15pm

Authors:

Palansooriya, K. N., Korea University
Wijesekara, H., Faculty of Applied Sciences, Sabaragamuwa University of Sri Lanka, Belihuloya, Sri Lanka
Bradney, L., The University of Newcastle
Kumarathilaka, P., University of Southern Queensland
Bundschuh, J., University of Southern Queensland
Bolan, N., University of Newcastle
OK, Y. S., Korea University

Particulate plastics (PPs) are emerging pollutants of increasing concern in aquatic and terrestrial environments. The occurrence and impact of aquatic PPs are extensively studied, but the studies on their occurrence in terrestrial ecosystems are mostly limited. This review work focused on the characteristics of PPs in terrestrial ecosystems for better understanding their behaviour and impact on terrestrial ecosystems. Important physical properties (i.e., size, colour, shape) and chemical properties (i.e., polymer type, associated chemical bonds, chemical additives) of PPs are discussed. In addition, details on biodegradable and biosynthetic PPs, and associated trace metals in PPs are discussed. Lastly, various analytical techniques that are used to extract, identify, characterize, and quantify (morphological and chemical characterization) the PPs in terrestrial environments are described.

This work was carried out with the support of "Cooperative Research Program for Agriculture Science and Technology Development (Effects of plastic mulch wastes on crop productivity and agro-environment, Project No. PJ014758)" Rural Development Administration, Republic of Korea. Corresponding author email: yongsikok@korea.ac.kr

Multi-Task Prediction of Hydrochar Properties from High-Moisture Waste with Machine Learning Methods

Conference: Waste Management

Year: 2020

Proceeding: Sustainable Waste Management Workshop

Group: General Submissions

Session: Session 2

Time: Wednesday, January 8, 2020 - 2:15pm-2:40pm

Authors:

Li, J., NUS
Wang, X., National University of Singapore

Hydrothermal carbonization (HTC) is a promising way to treat sludge, food waste, and manure with hydrochar production due to its low energy intake and high efficiency. Hydrochar with excellent fuel property is expected to be a promising fuel to replace fossil fuels for heat and electricity generation. Meanwhile, it can be used as fertilizer, conditioner or remediation agent in soils as a carbon-rich porous material. The char with stable carbon could provide a mean for carbon sequestration in soils. The traditional way to understand these important characteristics of hydrochar is to conduct HTC experiments and then detect the characteristics individually, which is quite expensive, labor-intensive and time-consuming. Machine learning (ML), as a data-driven approach, can perform prediction tasks after training with HTC dataset and facilitate understanding of the relative importance of input features. In this work, three state-of-the-art ML models including supporting vector machine (SVM), random forest (RF) and deep neural network (DNN) were used to predict the characteristics of hydrochar. The average R^2 of single and multi-task prediction of the optimized DNN model were 0.87 and 0.91, which indicated that the multi-task prediction performance of DNN was the best compared to other two ML models. This prediction can also help us make some evaluations about the hydrothermal technology and the application of hydrochar before taking action, which is beneficial to the labor, time, energy and resources saving compared with the experimental process.

Co-Gasification of Microplastics with Biomass for Syngas Production – a Simulation Analysis

Conference: Waste Management

Year: 2020

Proceeding: Sustainable Waste Management Workshop

Group: General Submissions

Session: Session 3

Time: Wednesday, January 8, 2020 - 2:30pm-2:55pm

Authors:

Suvarna, M., National University of Singapore
Wang, X., National University of Singapore

Co-gasification of microplastics with waste biomass to produce syngas serves as an effective means, to utilize biomass residue and mitigate the microplastics threat; while simultaneously producing energy. In this study, a steady-state simulation model for co-gasification of microplastics with biomass as the feedstock is developed using Aspen Plus. The study investigates the role of parameters such as microplastics to biomass ratio (M:B) in the feed, operating temperatures, equivalence ratio (ER) and steam to feed ratio (SFR) on the resulting syngas composition and its lower heating value (LHV). Preliminary results suggest that, in presence of O_2 as the gasifying agent, increasing the M:B ratio in the feed, gasification temperature and SFR increase the concentration of H_2 and CO in the syngas stream as well as its LHV. ER has an antagonistic effect as its increase hampers the production of H_2 and CO in the syngas stream and the corresponding LHV. This study provides a qualitative support for co-gasification of microplastics with biomass to achieve final syngas with content (H_2 + CO) greater than 62 mol%, H_2/CO molar ratio in range 0.81 â€“ 0.96 and LHV in the range between 7.2-8.9 MJ/Nm3 for the various operational conditions investigated.

Effect of Plastic Mulch Wastes on Soil Quality and Crop Productivity in Agro-Environments

Conference: Waste Management

Year: 2020

Proceeding: Sustainable Waste Management Workshop

Group: General Submissions

Session: Session 3

Time: Wednesday, January 8, 2020 - 2:55pm-3:20pm

Authors:

Dissanayake, P. D., Korea University
Kim, S., Korea University
Ok, Y. S., Korea University

*Corresponding author: Tel: +82-2-3290-3044, E-mail:yongsikok@korea.ac.kr

During recent decades, plastic production has increased significantly due to their easiness for use, cost effectiveness and durability. Nevertheless, plastic debris remained in natural environment has raised an emerging concern throughout the world. Plastic mulching has become a globally applied agricultural practice for its instant economic benefits such as higher yields, weed control, increased water-use efficiency and reduction of soil erosion. Despite of the short term benefits of plastic mulches, long-term consequences in terms of potential for deteriorating soil quality or their trophic transfer are less explored. This study is aimed to evaluate the effect of plastic mulch wastes on soil quality and crop productivity in agro-environments. First, plastic contamination in agricultural soils will be characterized and quantified in three agricultural sites in Korea and also soil properties and plastic degradability will be assessed. Based on environmentally relevant plastic concentrations derived from previous monitoring results, microcosm experiments will be designed and conducted to evaluate long-lasting fate of plastic mulch wastes in agricultural soil. In addition, recommended crops will be grown under same conditions and their growth will be assessed.

This work was carried out with the support of "Cooperative Research Program for Agriculture Science and Technology Development (Effects of plastic mulch wastes on crop productivity and agro-environment, Project No. PJ014758)" Rural Development Administration, Republic of Korea.

A Data-Driven Method for Constructing the Geography of Waste in Singapore

Conference: Waste Management

Year: 2020

Proceeding: Sustainable Waste Management Workshop

Group: General Submissions

Session: Session 3

Time: Wednesday, January 8, 2020 - 3:20pm-3:45pm

Authors:

Li, L., National University of Singapore
Wang, C. H., National University of Singapore
Wang, X., National University of Singapore

A substantial amount of microplastics are derived from the fragmentation of plastic waste. Robust planning of the waste management system requires reliable information on waste generation. It is an essential step to forecast the temporal-spatial variations of waste in the region for a given timespan. However, no study has covered the forecasting of spatial distribution of the subcategories of municipal solid waste, which are essential when designing the specific waste-to-resource/energy pathways for a circular economy. In the case of micro-plastic waste, it is important to know how the microplastic waste geographical distributes in order to better control it from source or remove it from the environment. To fill the research gap, this project focuses on developing a data-driven method for the prediction of future waste production and mapping the geography of waste for different waste categories. A â €œGeography of Wasteâ€ (GoW) approach containing the steps of data collection and analysis, visualization, data-driven modelling, and waste prediction is proposed in this work. It is applicable to any type of wastes including microplastics. With the retrieved dataset, different machine learning algorithms including neural network and gradient boosting are employed to identify how significance each factor affects the waste generation compared with linear interpolation. Results show that the gradient boosting model produces a reasonable prediction on the test dataset. In future work, the model will be applied to study the geography of microplastics, and the prediction models will be integrated with the decision support tools for the planning of waste management facilities.

Sensitive Method for the Detection of Micro Plastics in Environmental Samples

Conference: Waste Management

Year: 2020

Proceeding: Sustainable Waste Management Workshop

Group: General Submissions

Session: Session 4

Time: Thursday, January 9, 2020 - 9:30am-9:55am

Authors:

Sathrugnan, K. Sr., Frontier Laboratories

Plastic pollution is increasingly receiving attention from scientific community and policy makers. Recent focus is much towards microplastics- tiny plastic pieces ranging from 5mm down to 1 nm in diameter. These microplastics are filling the seas and working their way in creatures. This means, ocean microplastics are entering our food chain and finally to our bodies. Similarly, there are reports, detecting microplastics in bottled water, canned food and indoor air dust. This is a widespread problem in microplastics research that makes it hard or impossible to compare studies [1]. It is mainly because we donâ€™t have any good method yet to analyze microplastics in different matrices. In this paper, we present a simple method for analyzing microplastics using pyrolysis coupled to Gas Chromatography-Mass Spectrometry (Py-GCMS). In the case of bottled water or other natural waters, there are no extraction required. Water sample are filtered with prebaked glass fiber filter and microplastics retained on the filter will be directly analyzed by Py-GCMS. The method suitability is demonstrated by standard addition method with plastic mix at low concentration levels. With the PYGCMS and suitable software, very low detection limits are achieved. The method has shown advantages in minimizing sample preparation steps for complex environmental matrices.

Waste-Plastic Bottle Derived Composite for Sorption of Oil

Conference: Waste Management

Year: 2020

Proceeding: Sustainable Waste Management Workshop

Group: General Submissions

Session: Session 4

Time: Thursday, January 9, 2020 - 9:55am-10:20am

Authors:

Mitra, D., National University of Singapo

The global plastic pollution is massive, and overcoming it is a serious challenge. Poly ethylene terephthalate (PET) accounts for 11.7% of the plastic waste. While studies are being carried out plastic recycling, those reported for PET bottles are limited. Oil spills in aquatic bodies are a huge threat to the environment and the ecosystem. Apart from oil spills, various industries generate massive volumes of oily wastewater that requires treatment to regenerate clean water either for reuse or discharge. Most commercially available oil sorbents are made of synthetic polymers like polypropylene and polyurethane which have high sorption capacities, but the use of these introduces tons of non-biodegradable plastic in to the ecosystem.

In this work, LCBW which is acid-leached carbonaceous solid waste from gasification of refinery bottoms, has been entrapped within PET to fabricate a novel composite oil sorbent material. The composites exhibited a good oil sorption capacity of 7-8 g/g for mineral, paraffin as well as crude oils. The sorbent could also remove mineral oil from an oil in water emulsion containing ~5000 mg/L oil. The results from this study showed that a 3D porous material fabricated by a low energy and facile process, entirely from used PET bottles and carbon waste from the refinery could be effectively used as a sorbent for oil spill containment or for removal of oil from o/w emulsions in the industries. Thus, a potential way of tackling waste management as well as fabricating an effective oil sorbent material is achieved.

Vector Transport of Antibiotics By Microplastics in Water

Conference: Waste Management

Year: 2020

Proceeding: Sustainable Waste Management Workshop

Group: General Submissions

Session: Session 4

Time: Thursday, January 9, 2020 - 10:20am-10:45am

Authors:

Atugoda, T., Postgraduate Institute of Science, University of Peradeniya, Peradeniya, Sri Lanka
Vithanage, M., Ecosphere Resilience Research Center, Faculty of Applied Sciences, University of Sri Jayewardenepura
Wijesekara, H., Faculty of Applied Sciences, Sabaragamuwa University of Sri Lanka, Belihuloya, Sri Lanka
Werellagama, D. R. I. B., Wellington Institute of Technology
Bolan, N., University of Newcastle

This study investigates the sorption behavior of Ciprofloxacin (CPX) antibiotic, which is a common human and veterinary drug, on polyethylene microplastics (PEMP) through batch sorption experiments. The effects of ionic strength and dissolved organic matter (DOM) were examined by introducing $NaNO_3$ and Humic acid (HA) in to the medium. The sorption of CPX with pH had a gradual increase reaching a maximum sorption at pH 6-7 and then decreased which is likely due to the speciation of CPX at different pH. In the presence of $NaNO_3$, the overall CPX sorption capacity of PEMP decreased as the ionic strength increased revealing that the sorption mechanism was dependent of hydrophobic and electrostatic interactions. Sorption capacity of CPX by PEMP decreased significantly in the presence of HA highlighting the influence of DOM in the aquatic environment towards the CPX sorption and mobilization by PEMP. The spectrum for CPX sorbed PEMP, peaks for C=O, C-F and C-O-C confirmed bound CPX on MP surface. The sorption kinetics model confirmed the parabolic diffusion model and the Elovich model in the presence of HA. The isotherm equilibrium data was well fitted for Hill and Dubinin-Radushkevich models indicating sorption was controlled by pore filling mechanism accompanied by Van der Waals interactions on a heterogeneous surface. Hence, PEMP may act as a potential vector to transport CPX in aquatic environment where the sorption is influenced primarily by the characteristics of the water system and the properties of the adsorbate.

Extraction of Cellulose Acetate from Cigarette Butts As Raw Material for Making Membranes

Conference: Waste Management

Year: 2020

Proceeding: Sustainable Waste Management Workshop

Group: General Submissions

Session: Session 4

Time: Thursday, January 9, 2020 - 10:45am-11:15am

Authors:

Hakim, S. M., Institute of Technology Bandung
Anjani, K. D. W., Institute of Technology Bandung

Cigarette butts are the biggest marine contaminant. Cigarette butts is made by Ca which is a nonbiodegradable substance. Cellulose acetate is a type of polymer that takes 2-18 years to decompose. On the other hand, cellulose acetate can be used as membrane's raw material for desalinatiom of sea water to overcome the water crisis.

Forward Osmosis membranes made from cellulose acetate is an economical tool for seawater desalination. Literature studies show that there is a method for extracting cellulose acetate from cigarette butts. Leaching is carried out by solid-liquid extraction and followed by cellulose acetate precipitation. 0.02% H_2SO_4 solution, ethanol, and diethyl ether are used for solid-liquid extraction. Cellulose acetate precipitation is done by adding acetone and water which are then decanted and dried by the vacuum filtering method. The membrane has dope solution of cellulose acetate dissolved in a mixture of acetone, IPA, and lactic acid with an area of 0,0044 m^2. The membrane performance determined has a flux of 1,3 $L.m^{-2}.hours^{-1}$ and a Natrium rejection percentage of 98%. $MgSO_4$ with a concentration of 1,25 M is used as a draw solution with a flat sheet membrane configuration. It takes 20% of cigarette butts waste in Indonesia to make a seawater desalination plant that produces 100,000 m^3 / day of clean water.

Development of Biodegradable Polycaprolactone Film Incorporated with Antimicrobial Agent Via Extrusion Process

Conference: Waste Management

Year: 2020

Proceeding: Sustainable Waste Management Workshop

Group: Poster Session

Session: Poster Session

Authors:

Han, J., Korea University

In the present study, polycaprolactone (PCL) composite films incorporated with various concentrations of grapefruit seed extract (GSE) as an antimicrobial agent were prepared using a twin-screw extruder. Physical characteristics as well as antimicrobial properties of the PCL/GSE composite films were analyzed. The results showed that the surface color of the films gradually changed with increasing GSE concentration. Fourier transform infrared spectra indicated no significant structural changes such as chemical bond formation between PCL and GSE. Thermal properties were slightly affected due to GSE incorporation. Crystallinity of the composite films decreased as the amount of GSE increased. *In vitro* analysis indicated that the antimicrobial activity of the PCL/GSE composite films increased as the GSE concentration increased, with a 5% concentration showing the strongest inhibitory activity against*Listeria monocytogenes*, with 5.8-log reduction in bacterial count. Application testing of the films was carried out for cheese packaging, and biodegradation of the samples was assessed via soil burial testing. Our findings confirmed the potential use of PCL/GSE composite films as biodegradable food packaging material with antimicrobial activity.

Occurrence of Contaminants in Drinking Water Sources and the Potential of Biochar for Water Quality Improvement

Conference: Waste Management

Year: 2020

Proceeding: Sustainable Waste Management Workshop

Group: Poster Session

Session: Poster Session

Authors:

Palansooriya, K. N., Korea University
Yang, Y., The Education University of Hong Kong
Tsang, Y. F., The Education University of Hong Kong
Sarkar, B., Department of Animal and Plant Sciences, The University of Sheffield
Hou, D., Tsinghua University
Cao, X., Shanghai Jiao Tong University
Meers, E., Ghent University
Rinklebe, J., University of Wuppertal
Kim, K. H., Hanyang University
OK, Y. S., Korea University

In this work, the occurrence of contaminants in drinking water sources was described in relation to their treatment options based on both conventional (e.g., coagulation-flocculation, sedimentation filtration, and chlorination) and advanced treatment techniques (e.g., membrane filtration, ozonation, and biofiltration). However, due to apparent drawbacks of these methods (e.g., formation of disinfection by-products (DBPs)), it is desirable to develop an alternative option for safe drinking water. In this respect, biochar is recognized as an effective candidate to resolve the limitations in treating common pollutants typically occurring in drinking water such as microbial contaminants, inorganic contaminants, heavy metals, volatile organic compounds (VOCs), pharmaceuticals and personal care products (PPCPs), and endocrine disrupting chemicals (EDCs). As biochar can exhibit different types of interactions with adsorbates, its sorption processes can be explained by diverse mechanisms, e.g., p-p electron donoracceptor interactions, complexation, precipitation, H-bonding, and electrostatic attraction. In light of the attractive features of biochar (e.g., enhanced sorption properties, cost-effectiveness, and environmentally friendly nature), we offer in-depth discussion on biochar-based water treatment technologies for large-scale water purification operation. The abstract was adopted from: Palansooriya, K.N., et al., 2019. Occurrence of contaminants in drinking water sources and the potential of biochar for water quality improvement: A review. Critical Reviews in Environmental Science and Technology, pp.1-63.

This work was carried out with the support of "Cooperative Research Program for Agriculture Science and Technology Development (Effects of plastic mulch wastes on crop productivity and agro-environment, Project No. PJ014758)" Rural Development Administration, Republic of Korea.Corresponding author email: yongsikok@korea.ac.kr

Role of Antibiotic Bound Microplastics on Pathogenic Microbes: A in Vitro Antimicrobial Assay By Well-Diffusion Technique

Conference: Waste Management

Year: 2020

Proceeding: Sustainable Waste Management Workshop

Group: Poster Session

Session: Poster Session

Authors:

Vithanage, M., Ecosphere Resilience Research Center, Faculty of Applied Sciences, University of Sri Jayewardenepura
Ramanayaka, S., Ecosphere Resilience Research Center, Faculty of Applied Sciences, University of Sri Jayewardenepura
Mandakini, H. T., University of Sri Jayewardenepura
Salim, N., University of Sri Jayewardenepura
OK, Y. S., Korea University

The objectives of the study is to assess the sorption behavior of three broad spectrum antibiotics; Ciprofloxacin (CPX) and Tetracycline (TC) to polyethylene microplastic (PEMP), and to observe the influence of antibiotic-sorbed PEMP on some pathogenic microbes at 1 g/L solid solution mixture under *in vitro* conditions. Antibiotic concentrations, 100 mg/L, were spiked to PEMP solid solution and kept for shaking at 100 rpm for 12 h. Two Gram negative (*Escherichia coli*-ATCC 25922,*Klebsiella pneumonia*-ATCC 13883) and Gram positive pathogenic bacteria (*Staphylococcus aureus*-ATCC 25923, *Bacillus cereus*-ATCC 11778) were used for the study. Sterile Mueller Hinton agar medium was used (20 ml/plate) for well-diffusion assay with positive and negative controls. Culture plates were incubated at 37 °C for 24 h and zones of inhibition were measured. Only a limited adsorption of antibiotics was observed by the PEMP; 28 and 13 mg/g for CPX and TC respectively. The TC and CPX loaded PEMP showed antibacterial activity against *S. aureus*, *B. cereus* and *E. coli*. In the case of *B. cereus,* a bacteriostatic effect was demonstrated by TC loaded PEMP. None of the antibiotics loaded PEMP indicated any inhibitory action against the *K. pneumonia* at the concentration tested although the positive control (TC) showed a bactericidal activity. The results confirm the role of PEMP as a vector to transport antibiotics to control common pathogenic bacteria in the aquatic environment. Further studies will be conducted to understand the effect of different environmental conditions and minimum inhibitory concentrations on pathogenic and common microbes.

Sorption and Desorption of Organic UV Filters Onto Microplastics in Single and Multi-Solute Systems

Conference: Waste Management

Year: 2020

Proceeding: Sustainable Waste Management Workshop

Group: Poster Session

Session: Poster Session

Authors:

Leung, K. S. Y., Hong Kong Baptist University

Sorption studies of organic pollutants by microplastics (MPs) in single-solute systems are well established in the literature. However, actual aquatic environments always contain a mixture of contaminants. Prediction of the fate and biological effects of MPs-mediated chemical exposure requires a better understanding of sorption-desorption processes of multiple organic contaminants by MPs. In this study, the altered sorption and desorption behaviors of individual organic UV filters (BP-3 and 4-MBC) in the presence of cosolutes (BP-3, 4-MBC, EHMC and OC) on two types of MPs (LDPE and PS) were examined. In most cases, co-occurrence of other organic UV filters appeared to have an antagonistic effect on the sorption of primary solute, which was consistent with trends found in previous studies. Nevertheless, the sorption uptake of 4-MBC as primary solute on PS was enhanced in the presence of cosolute(s), arising presumably from solute multilayer formation caused by laterally attractive Ï€-Ï€ interactions between adsorbed cosolute(s) and 4-MBC molecules. Such formation of multilayer sorption in multi-solute systems depends on the solute hydrophobicity and concentration as well as inherent sorptivity of MPs. Our further desorption experiments revealed that the bioaccessibility of primary solute was significantly elevated with cosolutes, even though competitive sorption was observed under the same experimental conditions. These findings supplement the current knowledge on sorption mechanisms and interactions of multiple organic contaminants on MPs, which are critical for a comprehensive environmental risk assessment of both MPs and hazardous anthropogenic contaminants in natural environments.

AUTHOR INDEX

AIChE
120 Wall Street, FL 23
New York, NY 10005-4020

ISBN 978-1-7138-0814-5

9 781713 808145

CURRAN ASSOCIATES INC.
proceedings.com